Technology Skills for Kids

— Mental Preparation —

© 2023 Michael Gorzka. All rights reserved.
www.technologyskillsforkids.com

Hello and welcome!

This is **book #1** in a series of picture books created to help children make productive use of their computers and other devices while staying safe online.

Tech Wizard Mike & Tabby

Due to space limitations (even in a series of 12 books), we have to paint these technology skills in very broad strokes so to speak.

Any words or phrases in bold orange are topics that will be explored further in this book's supplemental materials.

NOTE: In this book and subsequent books in this series, we will use the word "device" as a **catch-all term** for computers, tablets, smartphones, wearables, multimedia boxes, mystery gadgets, et al.

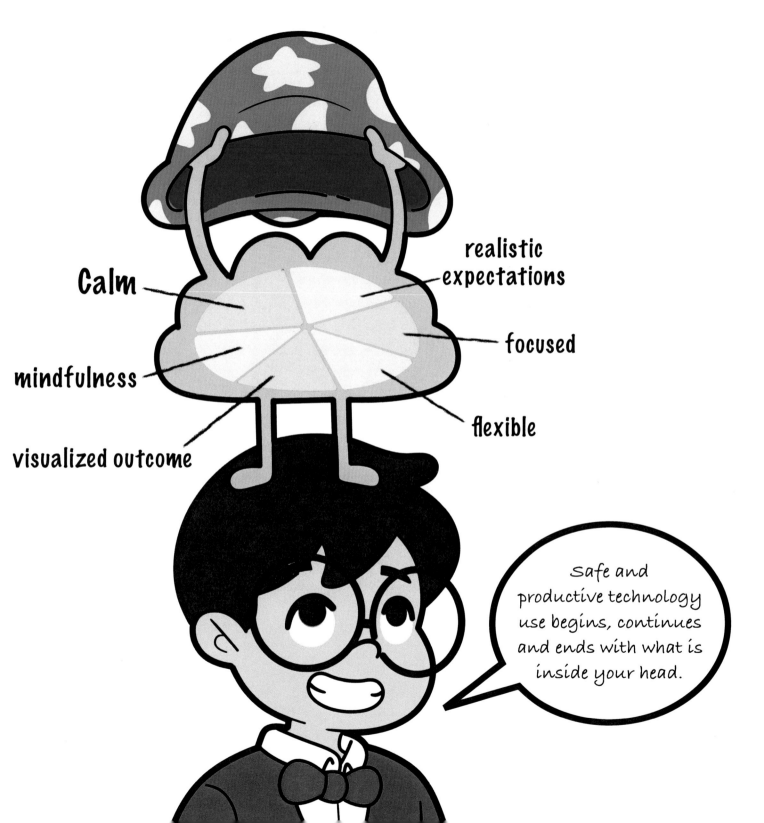

For example, you do **not** want to send any messages or post on social media while you are angry or upset — rather you want to be calm, cool and collected.

Cool as a Cucumber

Context matters — as much in the technology world as it does in the real world.

For example, you (probably) would not want to sing your favorite aria from La Bohème while standing in line at the supermarket.

Bee Careful!

The Internet (and social media in particular) has created many new situations for us to be bullied and to be **perceived** as being bullies.

Contemporary technologies (including the Internet) can make it easy to **get stuff done**…

(Responsible and effective technology use is not a result of age — it is a result of character.)

…but these same technologies can also make it easier to participate in aggressive and even hostile interactions. **Do not engage**. If something makes you feel uncomfortable, get some good help.

Exit from any abusive situations (online or offline) as quickly as you can.

We all need to be **very** careful with what we post, share and send.

Carol: "Pssst! You are not going to believe what I just did!"

Tech Wizard Mike: "Pssst! Try me!"

Actions often have reactions...

...in the real world **and** in the technology world.

Negative consequences (such as a damaged reputation) can result from our **acting without thinking** — technology makes it very easy to do this!

Our devices only know what we tell them…

…so we must be very careful what we tell them!

Our devices are all as dumb as a box of rocks...

...or a bag of hammers (pick your analogy).

"GIGO" STANDS FOR: GARBAGE IN, GARBAGE OUT

GIGO means you will get out of your devices what you put into them.

(This illustration shows a metaphorical outcome not a literal one 🙂)

For example, willy-nilly **computer mouse-clicking** will not get you anywhere (except lost in the weeds).

We should use our devices with as much thought and intention as we would use with any household appliance.

(Or who knows what could happen?)

Many **adult** technology users do not understand this concept!

If **you** get it, here is a gold star ⭐ Good job 👍

We have to expect the unexpected during technology use (just as we need to do in real life).

Visual analysis will always be very helpful. For example, look how friendly "Nessie" looks!

We need to apply what we have learned from living in the real world to our technology use.

For example, we need to be very alert and follow visual cues when **crossing the street**.

And also (just like in real life), it pays to be **organized**.

Many devices (e.g. computers, tablets & smartphones) will have a storage system that is similar (at least conceptually) to a **filing cabinet**.

Patience, persistence & practice are just as important in the technology world as they are in the real world.

The **crow pose** is a great analogy for success (both online & offline).

How to Make the Perfect Cup of Tea

(Yes, this is a technology skills book but please bear with me here.)

1. Start with fresh, cold water…
2. Place a tea bag in your favorite cup or mug.
3. Bring water to a rolling boil and immediately pour over your tea bag.
4. Steep for a good 3 to 5 minutes…
5. Remove the tea bag, relax and enjoy!

Question: What does brewing a perfect cup of tea have to do with becoming comfortable and proficient with technology?

Answer: Everything, really.

When you brew a cup of tea, you have a desired goal, an **objective**, a specific outcome in mind.

And to realize this objective, you methodically and mindfully follow a sequence of well-reasoned steps.

For example, you cannot pour the tea, add whatever you like to it and **then** heat the water — it wouldn't make logical sense to do so 🤔

Successful tea brewing also involves **patience** (e.g. 3 - 5 minutes for steeping) and **realistic expectations** (e.g. your tea will not be **hot** unless you are willing to heat some water ♨️).

Please approach **getting stuff done with technology** as you would brew a cup of tea.

"A cup of tea shared with a friend is happiness tasted and time well spent."

Seek out positive collaborations and companions.

You will have a sense of belonging when you are with true friends.

Share your friends' interests. Always be polite and attentive.

Caroline gave a very nice presentation on how to brew the perfect cup of tea.

Too much technology use can make for some **very** lame parties.

There are times and places for gadget use — a birthday party isn't one of them.

Expensive devices come and go…

...but shared moments with real-life friends are priceless!

There are a **plethora** of computers, tablets, smartphones, audio-visual contraptions and various gadgets of every stripe.

And there is a **plethora** of things you can do with any of those shiny gadgets.

And there is a **plethora** of things that can go awry if you do not use said gadgets with thought and intention.

Good Grief!

Due to the **plethora** of technology-related options available to us, it is very easy to lose sight of the forest through the trees, so to speak.

In order to **avoid** losing sight of the forest through the trees, we need to KISS our technology use.

But not this **way!**
(and do not forget about the things that really matter!)

KISS stands for "Keep it simple, snookums"

KISSing your technology use does **not** mean this!

Rule of thumb: Take only the technology-related "stuff" (e.g. computers, mobile devices, applications, peripherals and online services) that you need to thoughtfully accomplish what you need to do…

…and then go out have some fun in the real world — with **real people** who make you feel good about yourself!

"It's not what we have in life, but who we have in our life that matters."

During technology use, it is very important to utilize real-world observations and common sense.

Visit **www.technologyskillsforkids.com** for more tech skills for kids including blog posts, videos **and** book #2 in this series in which we will talk about graphical user interfaces.